Engineering in Organizations

+

Maintenance in Manufacturing

2 Books in 1

Engineering in Organizations

A Basic Introduction to the Mechanical, Electrical, Chemical, and Civil Branches

Nathan Brusselli and Louis Bevoc

Published by
NutriNiche System LLC

Mechanical Engineering	4
Electrical Engineering	13
Chemical Engineering	21
Civil Engineering	29

Mechanical Engineering

Introduction 5
Education 5
Natural ability 6
Responsibilities 7
Research and development 7
Equipment and machinery design 8
Equipment and machinery modification 8
Process design 9
Training 9
Project management 9
Work environment 9
Manufacturers 9
Job Shops 9
Laboratories 10
Offices 10
Travel 10
Temporary assignments 10
Future 11
General 11
Specific 12

Introduction

Engineering is the application of math and science to design, build, utilize, and maintain a wide variety of equipment, machinery, systems, and processes in organizations. It is a very broad field that can be broken down into many different subcategories that deal with specific aspects of the discipline. However, most experts break engineering down into four main branches for simplification purposes. These branches are mechanical engineering, electrical engineering, chemical engineering, and civil engineering.

Mechanical engineering concerns the construction and running of machinery and it will be the first branch discussed in this book. A mechanical engineer (ME) performs a variety of different tasks at work, but his or her major job function is defined as:

> The design, manufacturing, testing, maintenance, and usage of mechanical systems.

The roots of mechanical engineering can be traced back thousands of years....even to the actions of people in ancient Greece (BC). However, it was Sir Isaac Newton and his Laws of Motion that brought mechanical engineering in the spotlight using physics and mathematics as support. Years late, the first professional society of mechanical engineers was formed in Europe and a separate branch of engineering was officially developed.

Modern day mechanical engineering involves physics, math, and science...not to mention good old fashioned common sense. People in this profession also have an understanding of thermodynamics, kinematics, mechanics, and electrical components. They utilize their knowledge to develop, test, and analyze many different types of machinery, equipment, processes, and systems in places such as manufacturing plants and research labs. Their work typically has an industrial application, and they are often involved with projects from conception to finish.

Top-notch MEs combine education with natural abilities for effectiveness. These qualifications are defined in more detail as follows:

Education

Most mechanical engineering jobs require a bachelor's degree in Mechanical Engineering. This is obtainable from many different universities, and it usually takes four to five years to complete. In the United States, this degree is accredited by the *Accreditation Board for Engineering and Technology* (ABET). The ABET regulates course requirements for all colleges and universities to ensure adequate training and equal standards; and approved programs are listed on their website.

Much of the coursework in bachelor's degree programs focuses on math and science related concepts. Math typically consists of higher level calculus and differential equations, and science courses revolve around physics...but chemistry is also significant. It is also important to note that the other three main branches of engineering (chemical, civil, and electrical) are part of a mechanical engineer's curriculum.

The following types of courses are typically taken be MEs in order to earn their bachelor's degrees:

- Calculus
- Chemistry
- Computer-aided design (CAD)
- Computer-aided manufacturing (CAM)
- Differential equations
- Engineering design
- Engineering technology
- Engineering theory
- Fluid dynamics
- Fluid mechanics
- Hydraulics
- Instrumentation
- Kinematics
- Linear algebra
- Machine design
- Physics
- Pneumatics

In addition to the course work, most degree programs require the completion of a project to graduate. This usually takes place in a business rather than an educational facility, and it is designed to provide students with real world experience and let them apply their problem solving skills.

Natural ability

Mechanical engineers' natural ability is often just as important as their education. In fact, MEs with highest natural ability typically turn out to be the most successful.

Natural ability includes:

Active listening

This involves listening to what people are saying rather than just hearing them say it. There is a difference because hearing often goes "in one ear and out the other" while listening processes information for well-thought responses. Active listening is often a skill that is in high demand and short supply because people get distracted and fail to hear what is being said to them. This natural ability is important for many employees in organizations, but it is critical for MEs in order to insure they have attained the correct information for problem solving.

Creativity

Some people think creativity is a natural ability that is only necessary for artists, musicians, poets, and writers. People in these professions need creativity…but so do engineers. This is due to the fact many problems are not easily resolved, and creativity is sometimes an important part of the solution. Mechanical engineers face a wide variety of machine related issues that need to be overcome, and creative ability helps them find workable answers to the problems they encounter.

Innovation

Innovation is the process of coming up with new ideas or improving a concept that already exists. Based on this, it is understandable that innovation is a desirable quality in every mechanical engineer. After all, part of their job is to design and construct mechanical systems, and this requires original thinking. Without the natural ability to be innovative, MEs will not perform at levels expected of them by leaders of organizations.

Mechanical aptitude

This is the most obvious natural ability that is important for MEs. Mechanical engineers have to understand how things work….especially the inner workings of machinery. Their mechanical aptitude helps them solve problems where formal education is inadequate. In fact, some mangers make natural mechanical aptitude a top priority when hiring MEs. In short, people who have no mechanical aptitude should probably look at working in fields other than mechanical engineering.

Now you have a basic understanding of mechanical engineering and the skills needed to perform well in the profession. Next, let's move into a section on specific responsibilities of MEs.

Responsibilities

Mechanical engineers have a variety of different responsibilities where they apply their education, natural ability, and work experience. This is understandable based on the relatively high wages these individuals are paid in exchange for their services. They are expected to provide a return-on-investment, and that return-on-investment starts with responsibility.

The following are specific responsibilities of mechanical engineers:

Research and development

This is often assigned to mechanical engineers employed in higher education, but research and development (R&D) is also a focal point in some companies. In fact, organizations that are heavily involved in manufacturing often have an entire department devoted to R&D. However, regardless of the setting, R&D is something that MEs are usually responsible for at some point in their careers.

In higher education, R&D mechanical engineers often apply theoretical concepts in laboratory settings. This leads to new discoveries and innovation that drives technology in business. However, the downside is the fact that not all laboratory findings have real world application.

In industry, MEs with R&D responsibilities usually come up with new ideas for machines and other mechanical systems. They start with an idea or concept and try to bring it from the laboratory to the production floor where it can be utilized to lower costs and raise efficiency.

Equipment and machinery design

This is one of the most common responsibilities of mechanical engineers. They design equipment and machinery with a variety of different factors in mind including quality, safety, cost, and efficiency. These factors are broken down as follows:

Quality

Quality refers to the specifications of products, equipment, or machinery. If ranges or tolerances are specified, then mechanical engineering personnel are responsible for meeting them.

Safety

This refers to safety of the personnel operating equipment or machinery. Various safeguards need to be put in place so operators do not get hurt. Without these safeguards in place, injuries can be very serious….and even fatal is some situations.

Cost

Cost refers to the money needed get the equipment or machinery from the design phase to the shop floor. Essentially, it the labor and materials required to make the project successful. Cost also refers to the time and effort required to run and maintain the machine once it is operational.

Efficiency

Mechanical engineers need to get involved to make sure the equipment or machinery is meeting productivity expectations. Efficiency is very important in production-based facilities…and that is why MEs are employed in manufacturing plants.

Equipment and machinery modification

As noted above, efficiency is a responsibility of mechanical engineers. If equipment and machinery are not meeting performance expectations, then they need to be modified. Sometimes this can be done in process, and other time it means going back to the drawing board. Either way, MEs are responsible for performance modifications.

Process design

Some people assume that a lack of efficiency is due to the design of equipment or machinery. This is sometimes true, but it is not always an accurate assumption. The problem can be the process itself...and mechanical engineers are responsible for improving that process. They make changes to increase efficiency and lower costs based on observations and calculations.

Training

It should not come as a surprise that employees need to be trained on various operational and safety aspects of the equipment and machinery they operate. When this need arises, mechanical engineers are often the trainers because they were involved with that equipment and machinery from conception to implementation. They understand what needs to be done for safe and efficient operation, and they can answer any employee questions.

Project management

Most mechanical engineers are responsible for some type of project management. They oversee new production lines being implemented, procedures being changed, or expansions of workspaces. Regardless of the type of project, they are responsible for seeing it through to completion and this involves the management of people and processes. In this capacity, they also work as a liaison with outside contractors to ensure the project is finished in a timely and efficient manner.

Now you understand some of the major responsibilities of mechanical engineers. This leads us to discuss the surroundings where they perform their jobs...otherwise known as their work environment.

Work environment

This section discusses the conditions under which mechanical engineers perform their jobs. Work environment warrants discussion because MEs do not always work out of an office. In fact, some working conditions are far from that of an office.

The following are all environments where mechanical engineers work:

Manufacturers

This type of operation involves production lines that are used to assemble products. It is an excellent environment for the skills of mechanical engineers because equipment and machinery are crucial to the manufacturer's success. MEs are involved in every aspect of the production process, and the pace is often fast and furious. Manufacturing environments are action packed and rewarding in terms of accomplishments...but they can also be stressful and lead to burnout.

Job Shops

This environment is similar to manufacturing due to the machinery involved, but there is typically no assembly line involved. Small numbers of custom made items are usually produced to pre-determined specifications. An order can go back and forth to various areas of the shop, and flexibility is more important that productivity.

Some mechanical engineers prefer job shops over manufacturers because employees are more skilled and typically more concerned about their jobs. They tend to treat their machines better because they take ownership of their jobs, and this means less repair or modification is needed. Jobs shops also tend to have less turnover than manufacturers, thereby reducing the need for operational and safety training that are sometimes conducted by MEs.

Laboratories

Laboratories are typically reserved for R&D and quality work performed by mechanical engineers. Laboratory MEs design new products and make sure machinery and equipment are meeting designated standards. Laboratories are generally lower pressure environments than production or job shop floors, and this is why some MEs prefer working in them. Laboratories are also desirable because they tend to be on the cutting edge of technology…but the downside is that they do not offer the excitement or challenges found in production facilities.

Offices

This refers to the traditional environment for many white-collar business people. Some mechanical engineers prefer doing their jobs from behind a desk because this works well for them. They miss out on the cutting edge technology offered in laboratories and the excitement of plant operations, but they are willing to sacrifice that for a nine-to- five work schedule where their main tools are a phone, computer, and calculator.

Some mechanical engineering work environments differ in location rather than type. This happens when MEs travel or take on temporary assignments, both of which are described below.

Travel

Mechanical engineers sometimes need to travel to locations where problems are occurring so they can resolve them. This is especially when MEs are employed by consulting firms, but it also applies to companies that operate facilities in multiple locations. MEs need to visit sites that need their services, and they stay as long as necessary….usually from one day to two weeks. Sometimes their services are needed on a periodic basis. For example, a mechanical engineer might need to travel to the bakery division of multi-facility food processor every two months to address a variety of issues.

Temporary assignments

Temporary assignments are similar to travel with the difference being length of time at the location. In this situation, mechanical engineers are required to find housing for periods that can last several months to several years. Most MEs have a goal of getting back to their home base, but some prefer temporary assignments due to the change of scenery that prevents work environments from becoming stagnant. This change also provides new ideas and concepts that MEs can use for future problem solving.

As you can see, work environments of mechanical engineers differ depending on their specific job. However, regardless of their position, they need to ready for change because it is going to happen. This leads us to the last section that looks at future changes that will take place in mechanical engineering.

Future

In general, the future looks bright for mechanical engineers. This is due to the fact that mechanical systems will always need to be designed, manufactured, tested, and maintained. ME skills will be needed for a variety of different applications due to the problems that need to be solved. However, as might be expected, MEs will face obstacles as they move forward. Some of the challenges they will encounter include:

General

The following are general challenges that mechanical engineers will encounter:

Technology

Technology is important for mechanical engineers, but it will be even more important in the future. This is due to the fact that global completion will increase, and technological advances will play a critical role in which organizations are the most successful. MEs will need to understand, acquire, and implement technology in order to perform at optimal levels internationally. Astute MEs will realize that this technology comes from a wide variety of business applications that differ from those they find most comfortable.

Environment

Public perception of organizations today is often based on the impact those organizations have on the environment…and this is going to intensify in the future. Mechanical engineers will need to design and employ mechanical systems with a greater respect for the natural world. Waste and pollutions will be minimized, and recycling will play a bigger role in decision making. In short, MEs will improve the image of their employers though environment responsibility.

Education

Mechanical engineers will need to continuously update their skills…regardless of their experience or educational background. Learning opportunities in the form of webinars and online courses will continue to grow, and MEs will be expected to capitalize on them. Employers will realize the value of education and they will gladly pay for these types of services.

Cost

> Virtually every organization is impacted by money...and money will play a role in the future of mechanical engineering. Lower costs are necessary for lean manufacturing; and this means savings on research, design, operation, and maintenance of equipment and machinery will be more important than ever.

Diversity

> As with most forms of engineering, mechanical engineering it is dominated by white males. This creates a homogeneous workforce that prevents some opportunities for growth. In a nutshell, gender and minorities will increase their presence as MEs, and businesses will benefit worldwide.

Specific

One major challenge that mechanical engineers will face in the future involves energy related issues. Alternative sources of energy will need to be incorporated into the design of machines and equipment so their operation is energy efficient and cost effective. MEs will need to be conscious of the fact that traditional energy is not endless, and this will lead them to think about renewable sources for every aspect of their jobs. This is similar to what the automotive industry went through in the 1970s, except cost will be a larger factor than it was for car manufacturers.

Now you have an understanding of mechanical engineering. Let's use the same basic format to discuss the second branch of engineering known as electrical engineering.

Electrical Engineering

Introduction	14
Education	14
Natural ability	15
Responsibilities	16
Research and development	16
Data analysis	16
Computer-assisted design	16
Equipment or machinery modification	17
Training	17
Project management	17
Work environment	17
Manufacturing	17
Buildings and maintenance	18
Laboratories	18
Lighting	18
Business owners	18
Consulting	18
Future	19
General	19
Specific	20

Introduction

Electrical engineering is the branch of engineering that deals with the application and study of electronics and electricity. An electrical engineer (EE) performs a variety of different tasks at work, but his or her major job function is defined as:

The design, development, and testing of electrical components and systems.

The roots of electrical engineering can be traced back to the 17th century; and by 1800, a simple version of the electronic battery was developed. However, it was not until the 19th century, after the invention of the telephone and electric power, that electrical engineering became field of study.

Modern day electrical engineering involves physics, math, computer programming, and electronics. People in this profession also understand thermodynamics, machine design, and systems engineering principles. They use this information to develop and analyze communication equipment, power generation machinery, and many different types of motors. They typically work in factories, laboratories, research facilities, and offices; and their work often has a manufacturing or construction based application.

Top performing electrical engineers combine formal education with natural abilities for maximum effectiveness. These qualifications are defined in more detail as follows:

Education

Most electrical engineering jobs require a bachelor's degree in Electrical Engineering. This is obtainable from many different universities, and it usually takes four to five years to complete. In the United States, this degree is accredited by the *Accreditation Board for Engineering and Technology* (ABET). The ABET regulates course requirements for all colleges and universities to ensure adequate training and equal standards; and approved programs are listed on their website.

Much of the coursework in bachelor's degree programs focuses on math, science, and electronics. Math typically consists of higher level calculus, linear algebra, and differential equations; and science courses revolve around physics. Electronics courses explore problem solving...and other areas of engineering (mechanical, software, and power) are also part of the curriculum.

The following are types of courses taken be MEs in order to earn their bachelor's degrees:

- Calculus
- Chemistry
- Computer engineering
- Computer programming
- Digital logic
- Electrical circuits
- Linear algebra
- Mechanical engineering

- Physics
- Power engineering
- Probability
- Signals and systems
- Thermodynamics

In addition to the course work, most degree programs require the completion of a project to graduate. This usually takes place in a business rather than an educational facility, and it is designed to provide students with real world experience and let them apply their problem solving skills.

Natural ability

Electrical engineers' natural ability is often just as important as their education. As is the case with many engineers, EEs with the highest natural ability often turn out to be the most successful.

Natural ability includes:

Organization

Electrical engineers work can be complex when dealing with aspects of their jobs that involves circuitry or computer programs. For this reason, organizational skills are important. EEs without organizational skills often find themselves in confusing situations that result in guesswork...and this means they fall short of reaching goals and objectives.

Teams

Electrical engineers are often part of teams, so understanding how teams function and how to behave as a team member are important. Good team members understand that they cannot be in charge of every aspect because differing ideas and viewpoints will not be shared. They also realize that they cannot be "social loafers" who sit back and let others do the work. EEs who are effective team members are typically the most successful.

Communication

Most electrical engineers have valuable thoughts and ideas. They are trained to think and rationalize in order to solve problems. However, EEs are not always good communicators, and they sometimes have difficulty explaining and conveying their thinking. EEs who are naturally good communicators are able to transfer their thoughts and ideas to others, and this leads to the accomplishment of goals and objectives.

Trouble-shooting

Good maintenance people need to be more than mechanically inclined. If they cannot define the problem, then their repair skills are rendered ineffective. This same thinking applies to electrical engineers because EEs need to be able to troubleshoot problems in order to find solutions. This ability comes natural to some people...and those people are usually the best EEs.

Now you have a basic understanding of electrical engineering and the abilities needed to perform well in the profession. Next, let's move into a section that focuses on specific EE responsibilities.

Responsibilities

Electrical engineers have a variety of job responsibilities that require them to find economical and timely solutions to problems with many different variables. They need to be able to apply theory, think rationally, perform multiple tasks, make decisions, and work with others. It is difficult to list every job duty of EEs, but their major responsibilities are as follows:

Research and development

R&D electrical engineers are often employed in higher education, but R&D is also conducted in organizations.

Automotive suppliers and manufacturers are good examples of companies that employ EEs in R&D. These companies often build multi-million dollar laboratories designed solely for EE research. In these labs, electrical engineers often use computer design software. They start with an idea or concept, build prototypes, prepare reports and other documentation, and present their findings to higher management.

In higher education, electrical engineers often apply theoretical concepts in laboratory settings. This leads to new discoveries and innovation that drives technology in business. However, the downside of laboratory R&D is the fact that not all findings have real world application.

Data analysis

This ranks at the top of responsibilities for electrical engineers because they need to be capable of evaluating data received from components and systems in order to apply their knowledge and make decisions. Data analysis helps EEs design programs that create new products or solve problems encountered by employees in other areas of the organization. This makes everyone's job easier, and it saves time and money. That being said, people who do not possess data analysis skills probably should not be electrical engineers.

Computer-assisted design

Electrical engineers often work on projects. As part of the process, they typically use computer-assisted design (CAD) to visualize their ideas. These blueprints are important because they shows flaws and prevent projects from being continued when problems shows that they will not work. CAD allows EEs to efficiently put their ideas on paper or a screen, and it is a great tool for implementing specifications and adhering to other project requirements.

Equipment or machinery modification

This is an important responsibility for electrical engineers because they understand what machinery and equipment are capable of doing from an electrical standpoint. They observe operators in action and ask them questions pertaining to their jobs. After processing this information, they apply their knowledge and modify equipment or machinery to correct the problem.

Training

Some electrical engineering functions, such as CAD work or data analysis, are done strictly by EEs. However, this does not exempt them from training duties. Employees need to be trained on operational and safety aspects of the machinery they operate, and EEs are the best instructors because they can answer any questions related to the subject matter.

Project management

Electrical engineers work on many different projects. They test products, research cost-effective solutions, record and interpret data, and communicate their findings to customers and higher management. Regardless of the type of project EEs are assigned, they are responsible for seeing it through to completion...and this involves managing people and processes in an efficient manner.

Now you understand some of the responsibilities electrical engineers are charged with by their employees. Next let's discuss the work environments where they perform their jobs.

Work environment

Electrical engineers work in a variety of different environments, and it would be difficult to list each and every one of those environments in the scope of this book. However, some of the more common types are listed below.

Manufacturing

This environment is typically production based where assembly of products takes place. Electrical engineers are involved in aspects of the production process that involve electricity or electrical components. Manufacturing environments are rewarding in terms of action and accomplishments, but they can also be stressful and lead to turnover.

Along the same lines, electrical engineers are also employed in power distribution environments. These environments are similar to manufacturing, but the focus is on distribution of power in controlled systems. Areas such as power flow, short circuit analysis, and system impact are examined by EEs as they research technology and design new products.

Buildings and maintenance

In this environment, electrical engineers have a multitude of responsibilities regarding the electrical aspects building construction. These jobs require strong technical skills in additional to being able to multi-task and communicate with others. EEs comfortable with project management are often the best fit for this type of work situation.

Heating and cooling is another maintenance environment that employs electrical engineers. Heating, air conditioning, and cooling involve high voltage alternating current (HVAC) that requires the skills of EEs. They make sure indoor air quality and temperatures are maintained at acceptable levels by getting involved with all electrical aspects of the installation, operation, and maintenance of HVAC Systems.

Laboratories

Electrical engineers can find employment in a laboratory setting doing R&D work. In this environment, they typically design new products and make sure existing products meet electrical standards. Laboratory environments are generally lower pressure environments than production or job shop floors, and this is why some EEs prefer working in them. Laboratories are also desirable because they tend to be on the cutting edge of technology. However, the downside is that they do not offer the excitement or challenges found in production facilities.

Lighting

This is likely the most obvious type of work environment for electrical engineers. Most people understand that lighting involves electricity, so it makes sense that EEs are often involved. These engineers often work on commercial sites, overseeing all aspects of the lighting...from design to installation. This type of setting is good for EEs who are creative because projects often evolve and change in order meet customer needs and code requirements.

Some electrical engineers are self-employed. In this capacity, they are business owners who sell their services to other organizations. EEs also work as consultants. Consultants differ from business owners offering EE services because they oversee the work rather than perform it. Below is a more detailed description of each.

Business owners

Electrical engineers who own businesses sell their skills to organizations that need them. They perform electrical engineering services for the organization that utilizes them, but they are paid as a company instead of an individual. This works well for some EEs because they are able to control their work schedule and they do not have to answer to a boss. However, the downside to owning their own business is the fact that they have to supply their own benefits and insurance.

Consulting

Electrical engineering consultants can work for themselves or a consulting agency. As noted above, the difference between a consultant and a business owner is the fact that consultants oversee the work being done rather than performing it themselves. They make suggestion to improve products and processes, instead of physically making the changes themselves. They often work with engineers employed by the company that hired them a consultant. In this sense, they tend to operate as project managers with the company engineers answering to them.

Base on the above, you can see that electrical engineers perform the functions of their jobs in a variety of different environments. This means they have options in their careers... and those options often provide motivation for being the best they can at work. This leads us to the last section that discusses the future of EEs.

Future

Electrical technology is ever-increasing, and understanding of it is necessary for organizations to function efficiently. This takes a solid educational base and detailed work experience in the real world. Few people combine those skills, and that is why there will always be a demand for the talents of electrical engineers. However, there will be challenges involved as those engineers move forward. These challenges include:

General

The following are general challenges that electrical engineers will encounter:

First it is important to note that the skills of electrical engineers will need to be continually updated. Those with the most education and experience will be employed in the best jobs. This means EEs will need to undergo constant training in order to keep abreast of new concepts and technology. However, it will also open the door for engineers of all nations to be employed by top-notch corporations...as long as they are willing to continuously improve. That being said, the following are general challenges that electrical engineers will encounter in the future:

Transportation

Transportation is important for virtually everyone in the world. They might not own or lease their own vehicle, but they still need to be able to get to different places. The need for better and more energy efficient transportation will be a problem in the future, and electric vehicles will be a part of the solution. Electrical engineering technology will improve existing transportation by making it cleaner, safer, and more affordable. This will provide huge opportunities for EEs...but it will also pressure them to continually come up with new ideas and concepts.

Genetics

Genetic engineering will need electrical engineers in order to progress. In fact, it will be almost completely reliant on the detection, analysis, and simulation skills of EEs. Electrical engineers will be at the forefront of altering genes for improving the biological capabilities of people and other living organisms. In the future, genetic engineering ranks as a major challenge for EEs.

Social responsibility

Electrical engineers will need to make decisions with humans, ethics, and nature in mind. Profit will not be the only motive behind EE actions, and their main concern will be the society at large. Honest claims about research findings will be more important, environmental impact will have major significance, and consumer safety will be at the top of the list when designing new products. There are workable solutions to being socially responsible, but those solutions will come with many challenges.

Medical

Diagnostic tools used by professional in the medical field will rely heavily on electrical engineering design and technology. This will highlight the significance of EEs, but it also means the health and safety of patients will become their responsibility. For example, heart pacemakers relying on electronical engineering technology will be able to save people's lives…but the malfunctioning of these units will have the potential to be deadly.

Diversity

Lack of diversity is a common challenge for all types of engineering, but it needs to be noted for EEs. In short, electrical engineering it is dominated by white males. This creates a homogeneous workforce that prevents some opportunities for growth. In the future, gender and minorities will increase their presence as EEs…and businesses will benefit immensely.

Specific

One major challenge that electrical engineers will face is the need to acquire skills from other branches of engineering. Students majoring in electrical engineering will need to minor in mechanical engineering, civil engineering, or chemical engineering in order to meet the demands of global completion. Technical knowledge will rise in importance as EEs develop products and resolve problems in organizations all over the world. Some EEs are already experiencing this type of change, but it will become even more significant in the future.

Now you have an understanding of electrical engineering. Let's use the same basic format to discuss the third branch of engineering known as chemical engineering.

Chemical Engineering

Introduction	22
Education	22
Natural ability	23
Responsibilities	24
Safety	24
Product development	24
Evaluation	24
Design	25
Environment	25
Training	25
Work environment	25
Offices	25
Manufacturing	25
Laboratories	26
Worksites	26
Consulting	26
Travel	26
Future	27
General	27
Specific	28

Introduction

Chemical engineering is the branch of engineering that uses science and math to solve problems related to food, chemicals, pharmaceuticals, and other products. The job function of a chemical engineer (CHE) is defined as:

The conversion of chemicals, energy, and materials into useful products or processes.

The first basic courses in chemical engineering were taught in the United States in the late 1800s. However, it was not until the mid-1900s that this branch of engineering gained recognition and was more clearly defined as a discipline.

Modern day chemical engineering involves chemistry, physics, microbiology, biochemistry, calculus, thermodynamics, and kinetics. CHE's knowledge of these subjects is used to solve a wide variety of problems associated with work related conditions, processes, and procedures. These engineers typically work in factories, offices, and laboratories.

Top-notch chemical engineers combine formal education with natural abilities for maximum effectiveness. These qualifications are defined in more detail as follows:

Education

Most chemical engineering jobs require a bachelor's degree in Chemical Engineering. This is obtainable from many different universities, and it usually takes four to five years to complete. In the United States, this degree is accredited by the *Accreditation Board for Engineering and Technology* (ABET). The ABET regulates course requirements for all college and universities to ensure adequate training and equal standards; and approved programs are listed on their website.

The coursework in bachelor's degree programs focuses on science, math, and engineering related subjects. Science typically consists of chemistry, physics, and biology; math revolves around higher level calculus; and engineering related subjects include thermodynamics, process design, and biotechnology.

The following are types of courses taken be CHEs in order to earn their bachelor's degrees:

- Biology
- Calculus
- Chemistry
- Computer engineering
- Data analysis
- Fluid Mechanics
- Heat transfer
- Kinetics
- Linear algebra
- Organic chemistry
- Physics

- Product design
- Thermodynamics

In addition to the course work, most degree programs require the completion of a project to graduate. This usually takes place in a business rather than an educational facility, and it is designed to provide students with real world experience that allows them to apply their problem solving skills.

Natural ability

Chemical engineers' natural ability is often just as important as their education. As is the case with many engineers, CHEs with the highest natural ability often turn out to be the most successful.

Natural ability includes:

Problem solving

The natural ability to solve problems is important for any engineer, but it is particularly important for chemical engineers because they are often addressing multiple problems at the same time. For example, CHEs addressing efficiency problems with processes in manufacturing hazardous chemicals also need to be concerned with worker safety and environmental impact. They must have the problem solving skills necessary to correct the issue, safeguard employees' physical well-being, and prevent any type of damage to the environment.

Resourcefulness

Chemical engineers need to be innovative and creative when they apply old techniques to new problems. These types of situation require them to think "outside the box" and custom design solutions. Unfortunately, solutions are not always obvious, so CHEs need to be clever and original...and this is when their natural ability to be resourceful comes in handy.

Methodical

A major part of problem solving is figuring out why something does not work as planned. This analysis requires a step-by-step process that eliminates some factors and brings others to the forefront. Questions need to be asked and trial-and-error often plays a big role. In short, systematic problem solving is important for CHEs...and those who possess the natural ability to be methodical find it very advantageous.

Logical

Chemical engineers must use sound reasoning when designing, analyzing, or troubleshooting situations. In other words, they need to take a logical approach to problem solving. This might seem rather simple, but it can be challenging when multiple variables are often involved. Fortunately, the ability to apply logical in problem solving can be learned...but those who naturally possess the trait will have a big head start.

Now you have a basic understanding of the natural abilities chemical engineers need to perform well in their profession. Next, let's move into a section on CHE's specific responsibilities.

Responsibilities

Chemical engineers perform a variety of different job tasks, and the responsibilities associated with those tasks can vary extensively from one engineer to another. For example, some CHEs work only with specific processes such as polymerization, while other focus on entire fields such as biomedical engineering. They are further divided within these areas because some CHEs do research while others work on manufacturing or process problems. Add this to the fact that CHEs work in many different industries (energy, food, plastics, paper, automotive, etc.), and it is rather obvious that listing every type of responsibility assigned to them would be very difficult. However, the following are some of their major duties:

Safety

Safety is a responsibility of many chemical engineers because they work in environments that manufacture, process, utilize, or handle hazardous chemicals. CHEs develop procedures for working with these chemicals based on their understanding of chemical reactions and the potential for injurious or lethal situations. If something is not safe, CHEs work to find alternative processes, raw materials, or chemistry to reduce the risk of injury. The importance of this responsibility should never be underestimated because it prevents humans from being harmed or, in extreme situations, killed.

Product development

Chemical engineers are often involved in product development. Interestingly, they sometimes find success by taking a backwards approach. They are well trained for finding more efficient ways of doing things, so their road to product development often starts with defining better processes. They identify workable solutions to process problems, and then work backwards to develop unique products that are safe and cost-effective. This is not a new concept for scientists and, in reality, CHEs make up some of the best scientists in the world.

Evaluation

Solutions to problems in manufacturing are often found by chemical engineers. They troubleshoot by evaluating situations with a "think before act" mentality. They analyze equipment and processes first, and then they make decisions based on their education and experience. This evaluation procedure is similar to that used by other engineers...with one difference. That difference is the fact that CHEs always find solutions to problems with safety and environmental concerns in mind. If the end result has safety violation or negative environmental impact, then CHEs work toward a better alternative.

Design

This refers to process design rather than product design. Process design is the centerpiece of chemical engineering because it extracts and utilizes all components in the field. Because of this, CHEs are often charged with developing new processes that resolve existing issues. They implement plans with procedures that coincide with equipment layout for maximum safety and efficiency. In terms of design, CHEs are usually the best employees that organizations have to offer.

Environment

As noted earlier, environmental concerns play a big role in chemical engineering. CHEs work to prevent environmental mishaps from occurring, and they also find solutions to environmental problems that already exist. They do this by implementing procedures that control pollution, reduce waste, and conserve energy. Mother nature never takes a backseat in a CHE's work, and this is what differentiates this field from other engineering types.

Training

Yes, training is again mentioned as a responsibility of a specific type of engineering. Chemical engineers usually conduct some type of training, and it is often geared toward safety. CHEs are responsible for the safety of personnel at all times, and they invest substantial time and effort into preventing work related injuries. Their training efforts are important...especially for employees who working with dangerous or hazardous chemicals.

Now you understand some of the basic responsibilities of chemical engineers. Let's move on to a discussion on the work environments where they perform their jobs.

Work environment

Surprisingly, many chemical engineers work along other white collar workers in traditional office settings...although this is not always the case. The following are specific environments where CHEs find employment:

Offices

As noted above, chemical engineers frequently work in offices. They do this simply because they are able to work behind a desk and perform their job functions. They write polices, programs, and procedures based on their expertise; and their specifications are implemented by processing or manufacturing personnel. Training might be required, but sometimes even that can be done in an office setting.

Manufacturing

Sometimes chemical engineers find employment in manufacturing processes. On rare occasions they act as supervisors who oversee production personnel, but they are usually hired to monitor processes. This process monitoring might be for efficiency reasons, but it is more often related to safety and environmental concerns. CHEs can work for just about any manufacturing plant, but they are typically found in chemical, food, clothing, and fuel production facilities.

Laboratories

Similar to other engineers, chemical engineers also work in laboratory settings. However, CHEs are probably the most likely type of engineer to find employment in this environment because a large part of their jobs involve research. They look for ways to improve products or processes, and much of this work can be done in laboratories. For example, they might be working on ways to improve the quality of food products...and this starts with laboratory testing and analysis. Laboratories also offer the opportunity to experiment with new technology that is not ready to be part of plant or production processes. Some CHEs prefer working with new technology over any other type of job function, and that a major reason why they chose to work in the field of chemical engineering.

Worksites

This refers to local worksites where chemical engineers find the need for their services. For example, many chemical spills require CHEs to spend hours, days, or weeks overseeing the cleanup. They understand the safety and environmental concerns involved, and they can prevent many other problems from occurring. In a situation involving a hazardous spill, CHEs are well worth the money spent for their expertise.

Consulting

Chemical engineers are also employed as consultants. In this role, they suggest ways to improve products or process while emphasizing the importance of safety and environment. For example, they might be hired by a pharmaceutical company to find ways to make a particular drug safer for human consumption. Their suggestions might cost a substantial amount of upfront money, but they prevent problems such as lawsuits from occurring in the future. For this reason, their expertise is highly valued by management in organizations.

Travel

Sometimes travel is required for chemical engineers. This is unavoidable if, for example, they need to oversee safety concerns at a worksite that is not local. Obviously, the worksite cannot be brought to the CHEs...so they need to travel to it. This travel can last from one day to two weeks or longer...and repeat visits might be necessary.

Now you have a basic understanding of chemical engineering. Let's move forward to the next section that looks at the future of CHEs.

Future

Chemical engineers will be in demand, but they will likely have less overall opportunities than other types of engineers because they tend to be more specialized. However, the upside to this is the fact that CHEs who find employment will be paid handsomely because of their specialized skills.

Some of the future challenges chemical engineers will face include:

General

The following are general areas where chemical engineers will face challenges:

Medical

New medicines will be needed to treat cancer and virus based diseases. Working with chemists and medical professionals, chemical engineers will be at the forefront of finding safe and effective drugs to fight these deadly illnesses. They will do this using their expertise to discover new materials and processes that change the current ways of thinking. The field of medicine offers many opportunities for CHEs, but those opportunities will be challenging because they will require change that takes people out of their comfort zones.

Biomedical

Chemical engineers will get involved with DNA, RNA, organs, and tissues in order to improve the longevity and quality of human lives. For example, they might conduct research to find ways to regenerate body organs, thereby reducing the need for donors. New products and procedures will be developed by CHEs that put biomedical ideas into action. Thoughts that were once only a dream will become reality...as long as CHEs are willing to invest the necessary time and effort.

Ecological

The environment has been a concern for many years, and that concern will intensify in the future. Chemical engineers understand the chemistry involved with the earth, water, and air; and they will use their knowledge to reduce pollution, waste, and other threats to the natural world. Ecological issues will present problems, but CHEs will be able to provide the solutions.

Economical

Cost is an area where future chemical engineers will have an opportunity to shine. For example, they can improve the living standards for millions of people using low cost energy solutions. Their research skills will lead to uses of solar energy that will be economically and environmentally beneficial. They will also work on new processes to harness wind and ocean currents for energy sources that have not been utilized to their potential. The challenge will be to produce the most amount of energy with the least amount of environmental impact at the lowest cost. This will be difficult, but it is achievable.

Specific

A major challenge that chemical engineers will face involves the climate. Climate change is a real threat, and future CHEs will need to address that threat using their expertise. This will require the implementation of intervention strategies designed to prevent further damage and reverse the harm that has already been done. It will also require the rebuilding of ecosystems starting with those that have had the most devastating impact. This challenge will be difficult, but it can be done...and it starts with CHEs.

Now that you have an understanding of chemical engineering, let's move on to the fourth and final branch of engineering known as civil engineering.

Civil Engineering

Introduction   30
   Education   30
   Natural ability   31

Responsibilities   32
   Construction engineering   32
   Geotechnical engineering   32
   Structural engineering   32
   Transportation engineering   32
   Surveying   32
   Planning   33
   Managing   33
   Reporting   33
   Testing   33
   Maintaining   33

Work environment   34
   Government   34
   Utilities   34
   Non-residential buildings   34
   Non-residential structures   34
   Mobile   34

Future   35
   General   35
   Specific   36

Introduction

Civil engineering is the last major branch of engineering that will be discussed in this book. It deals with the overseeing of physical structures including buildings, roads, highways, bridges, tunnels, pipelines, and sewer systems. For simplification proposes, the job function of a civil engineer (CVE) is defined as follows:

> *The planning, design, analysis, and implementation of infrastructure in urban and rural areas.*

The roots of civil engineering can be traced back to the 1700s. At that time, military engineers were working on projects specifically designed for defense. Other engineers were working on non-military projects, and both types needed a way to be distinguished...so the term "civil engineer" was established. Interestingly, many of the functions performed by military engineers of the past have been taken over by civil engineers. This makes CVEs the most important of all engineers in terms of infrastructure development.

Present day civil engineering utilizes math, science, and other engineering for problem solving. Math involves calculus, linear algebra, statistics, and differential equations; science revolves around physics, but geology and chemistry also play; and the other engineering disciplines consist mainly of structural, materials, and environmental. A CVE's work typically involves construction of non-residential buildings or other structural projects that focus on the transportation of people, solids, liquids, or gases.

The best CVEs combine education with natural abilities for effectiveness. These qualifications are defined in more detail as follows:

Education

Most civil engineering jobs require a bachelor's degree in Civil Engineering. This is obtainable from many different universities, and it usually takes four to five years to complete. In the United States, this degree is accredited by the *Accreditation Board for Engineering and Technology* (ABET). The ABET regulates course requirements for all colleges and universities to ensure adequate training and equal standards; and approved programs are listed on their website.

The following types of courses are typically taken be CVEs in order to earn their bachelor's degrees:

- Calculus
- Chemistry
- Computer-aided design (CAD)
- Computer engineering
- Differential equations
- Fluids and heat transfer
- Geology
- Hydraulics
- Linear algebra
- Physics

- Structural analysis
- Surveying
- Transportation engineering
- Thermodynamics

In addition to the course work, most civil engineering degree programs require the completion of a project to graduate. This usually takes place in a business rather than an educational facility, and it is designed to provide students with real world experience and let them apply their problem solving skills.

Natural ability

Civil engineers' natural ability is often just as important as their education, and those with the highest natural ability typically turn out to be the most successful.

Natural ability includes:

Communication

Civil engineers need to explain their ideas and plans to a variety of different people in work related situations. Based on this, it is understandable that clarity is important for CVEs. They need to be able to clearly communicate with others in order to avoid the mishaps and problems that prevent projects from being successful. This natural ability is important for all engineers, but it is critical for CVEs to ensure problems are solved projects are properly completed.

Critical thinking

Civil engineers face a wide variety of problems that need workable solutions. In order to be successful, they need to gather and process information. They do this by combining known concepts with sound reasoning. This allows them to formulate conclusions while assessing strengths and weaknesses of those conclusions. In short, CVEs need to properly evaluate situations...and critical thinking is the core of evaluation.

Decision making

Not surprisingly, civil engineers are often in charge of projects. They need to keep those projects moving forward while making sure applicable codes, rules, regulations, and standards are being upheld. This is accomplished by telling people the things that need to be done and instructing them on how to do those things. Based on this, it is understandable that decision making is a desirable trait for CVEs. Without the natural ability to make decisions, CVEs will not perform at levels expected of them in their roles as project managers.

Vision

This might be the most important natural ability for civil engineers. CVEs often oversee projects from conception to completion, and this requires them to visualize what needs to be done. Vision is something that cannot be obtained through formal education. It is a natural ability, and people who do not possess it should likely look at working in fields other than civil engineering.

Now you have a basic understanding of the education and abilities necessary for civil engineers. Next, let's move into a section on the specific responsibilities of CVEs.

Responsibilities

Before going on to specific responsibilities, it is important to note that civil engineers usually specialize in one of the following areas:

Construction engineering

These engineers deal with the planning, execution, and management of construction and other infrastructure such as tunnels, roads, buildings and utilities. These engineers combine the skills of civil engineers and construction supervisors so they can see projects through to completion.

Geotechnical engineering

These engineers are concerned with materials in the earth (dirt, rocks, minerals, etc.) and the construction that occurs on those materials. They make sure foundations are solid, redirect water flow, and develop retaining walls for specific applications. Essentially, civil engineers focus on structures, and geotechnical engineers focus on the support for those structures.

Structural engineering

These engineers calculate the stability and durability of structures such as building, bridges, dams, or sewer systems. They are skilled in structural design, and understand the requirements necessary for safety. Their work is often conducted before any part of the structure is built, but their expertise is also required throughout the project.

Transportation engineering

These engineers apply science to the establishment and management of projects involving transportation. More specifically, they focus on the safety and convenience of transportation systems by overseeing the engineering aspects of construction. Major projects include the construction of streets, highways, railways, airports, street cars, and airports.

The above specializations can overlap in some situations, but they need to be understood as divisions of civil engineering in order to get a clearer picture of CVE responsibilities. These responsibilities are as follows:

Surveying

This is likely the most well-known duty of civil engineers. In this capacity, they access and interpret land and geographically information from the job sites where they are working. More specifically, CVEs are often responsible for grades, elevations, and reference points necessary for guiding the building or construction process.

Planning

This responsibility involves the planning of systems and structures. Planning is necessary because standards need to be adhered to and requirements need to be met. It starts with a vision that typically utilizes computer-assisted design (CAD) to implement project specifications and indicate flaws that could lead to failure. Once the vision is found to be acceptable, the plan is able to move into the construction stage.

Managing

Civil engineers are often responsible for construction projects. This involves overseeing people and processes, and it means that CVEs need to manage. Without management, projects move in many different directions and goals fail to be achieved. CVEs often make the best project managers because they understand what needs to be accomplished and the most efficient path to those accomplishments. In short, their expertise is used for decision making that leads to objective achievement.

Reporting

Civil engineers are often assigned the reporting duties. For example, they need to conduct site investigations and report findings. They also need to research ergonomic, economic, and environmental concerns and report discoveries. Additionally, they need to examine proposals, bids, deeds, and leans to report discrepancies. These reports are important for any type of building or construction...and they are the responsibly of CVEs.

Testing

This is essentially a form of research and development, but it specifically involves the assessment of materials used in construction. Civil engineers understand which building materials are best for jobs based on their expertise, and they test those materials to make sure they meet specific requirements. In short, CVEs are assigned the responsibility of making sure materials are safe, meet code, and will do what they are designed to do.

Maintaining

All construction must be maintained. It might be years before a structure needs to be serviced or repaired, but it will happen at some point...and someone needs to be responsible. Civil engineers oversee maintenance projects such as the repair or replacement bridges, dams, highways, pipelines, tunnels, roads, sewers, and other infrastructure. This responsibility is important because it keeps construction effective, safe, efficient, and modern.

Now you understand some of the major responsibilities of civil engineers, so let's move forward to the next section that discusses their work environment.

Work environment

Work environments for civil engineers vary depending on their career choices. For example, a CVE who works for a city government will likely experience a vastly different environment than a CVE who builds highways in foreign nations. However, these individuals will share common job functions.

The following are some of the more common work environments for civil engineers:

Government

Civil engineers often find work in state, local, or federal government agencies. In this capacity, they work in offices or offsite, and their jobs can be somewhat diverse. For example, they might survey land for a certain period of time and then get involved with the building of government offices. This type of work environment is generally lower pressure than that in private industry, and that is why some CVEs prefer it.

Utilities

Utilities are chiefly made up of gas, water, and electricity. Civil engineers find employment in these environments in order to oversee projects involving sewers, pipelines, cables, and wires. They understand the materials being used and the impact those materials have on the earth. In short, utility companies rely on civil engineers for safety and efficiency.

Non-residential buildings

Office buildings, exhibit halls, and stadiums are all examples of this type of work environment. The work for non-residential buildings often starts outdoors with the foundation and finishes indoors with structural details involving doors, stairways, and windows. Civil engineers are needed for the design and construction stages because they understand buildings from a practical, environmental, and safety aspect.

Non-residential structures

Statues, monuments, and artistic creations are examples of non-residential structures that civil engineers are involved with building. This work environment can be outdoors or indoors depending on needs or specifications. Similar to non-residential buildings, CVEs are involved with the foundation and structural details. However, many of these structures are not meant to have people inside, so safety tends to be less of a concern.

Mobile

Mobile work environments require civil engineers to travel with projects unit they are completed. Examples include roads, highways, and railways. The upside to this type of work is the fact that CVEs always get a change of environment, but the downside is that projects can go on for long periods of time with many days spent on the road traveling.

As you can see, the work environments of civil engineers vary extensively from job to job. They can work indoors, outdoors, underground, or on the water. However, regardless of the environment, all CVEs perform similar job functions. Now let's move on to the future of this profession.

Future

Building and construction will likely never end, and because of this there will be many opportunities for civil engineers in the future. However, CVEs will not be without challenges as they move forward. Some of these challenges include:

General

General areas where civil engineers will face challenges include:

Political

Like it or not, civil engineers will need to be more political in the future. They will become more involved with the environment and infrastructure, and this will force them to become more involved in the related issues. CVEs will be on committees that establish policies for environmental protection, encroachment, and public safety. They will need to become astutely aware of the legal aspects of building and construction, and this will be challenging for CVEs who want nothing to do with politics.

Safety

Safety is already important for civil engineers, but that importance will increase in the future. People will want to know that the buildings and structures they utilize are safe, and CVEs will provide that assurance. If CVEs fail to raise the bar on safety, then injuries and lawsuits will result.

Resilience

This refers to the resilience of the buildings constructed. Civil engineers will need to make buildings better and stronger so they can withstand natural disasters such as tsunamis, earthquakes, and tornadoes. Obviously, complete resilience will never be achievable, but the goal will be to constantly improve.

Congestion

This refers to traffic congestion. More vehicles will be on roads in the future, and civil engineers will be charged with making sure traffic does not get out of hand. They will need to design more efficient roads, highways, tunnels, and bridges so people can get where they need to be in a reasonable amount of time. This will take planning because space is not always available, but it can be done…and CVEs will be responsible for leading the way.

Specific

A major future challenge faced by civil engineers involves pollution. More specifically, CVEs will need to find ways to stop environmental pollution. They will need to utilize alternative sources of energy that are less damaging to the environment. They will also need to protect the earth that structures are built on and prevent ground water from becoming contaminated. Humans often pollute environments in the name of progress…and CVEs will need to formulate methods for preventing some or all of that pollution.

Maintenance Programs in Manufacturing
An Introduction to Preventative, Predictive, and Corrective Types

Louis Bevoc

Published by
NutriNiche System LLC

Introduction ... 39
Assembly and disassembly ... 39
Machine repair ... 39
Liaison ... 39
Testing ... 40
Calibration ... 40
Planning ... 40
Who will be involved? ... 41
What machinery will be involved? ... 41
What are the priorities? ... 41
What will be documented? ... 41
What is the operating budget? ... 42
Preventative maintenance ... 42
Advantages ... 43
Disadvantages ... 44
Predictive maintenance ... 45
Advantages ... 45
Disadvantages ... 46
Corrective maintenance ... 47
Advantages ... 47
Disadvantages ... 48
Summary of the three types ... 49
Summary ... 50

Introduction

Welcome to a world that is vastly underestimated in value. It is a world where "good morning" is replaced with "I have a machine down." It is a world where employees work before, during, and after production activities. It is a world where criticism is frequent and praise is rare. It is the world known as maintenance in manufacturing.

The above paragraph might be a little dramatic, but it is reality in many manufacturing facilities. Maintenance keeps production lines operating so orders can be filled and customers will be happy. In this regard, maintenance is one of the most important aspects of a production facility. That being said, what are the specific duties of maintenance personnel? The following is a list of their major job responsibilities:

Assembly and disassembly

When new machinery enters the facility, maintenance people are the first to have contact with it. They uncrate it, assemble it, and set it up. They are responsible for making sure all parts are included and the machine will do what it is supposed to do.

Maintenance people also disassemble machinery when it needs to be broken down for cleaning, repair, or removal. They understand the inner workings of the machines, and they know how to properly dismantle them without damaging them or creating safety risks for themselves or others.

In short, the buck stops with maintenance people in terms of assembly and disassembly.

Machine repair

Machines break down in manufacturing facilities. This happened in the past, it happens now, and it will happen in the future. However, broken equipment is not a major issue if it can be fixed...and maintenance people fill the role of fixers. They understand what needs to be done in order to get machines running properly in a reasonable amount of time. This saves manufacturers money, and it allows them to get their products to the customers who need them. In terms or machine repair, maintenance people are worth their weight in gold.

In short, machines need to be repaired and maintenance people fill that need.

Liaison

When machinery cannot be repaired, it needs to be serviced by external professionals. This means a technical person from the manufacturer of the machine needs to visit the facility to make the necessary repairs. Someone needs to be responsible for contacting that technical person and describing the problems the machine is experiencing...and that someone is almost always a maintenance person.

Describing machinery problems might seem like a relatively simple task. After all, the machine is broken, so what else needs to be said? Unfortunately, a lot more usually needs to be said. Technical people need detailed information in order to make a timely and proper diagnosis, and that information is only available from those who have a working knowledge of the machine. Typically, the only individuals with that working knowledge are maintenance people.

In short, maintenance people are liaisons who reduce the time, effort, and expense required for external repair of machinery.

Testing

How do production people know if a machine is able to meet their expectations? The answer is through testing. For example, a light intensity machine at a flashlight manufacturing company needs to measure the bulb brightness of products made on a high speed production line. One bulb needs to be measured every six seconds in order for plant personnel to meet established production quotas. Every day before production begins; maintenance people run tests to make the light intensity machine can handle the volume.

In short, maintenance people verify machines are capable of doing what they are supposed to be doing.

Calibration

How do production people know that a machine is working properly? The answer is through calibration. For example, a scale in a meat processing plant weights weighs one pound packages of hot dogs. It is definitely capable of weighing these hot dogs, but is it producing accurate results? The only way to find out is by calibrating the scale with standard weights...and this is done by maintenance people.

In short, maintenance people verify machines are accurately doing what they are supposed to be doing.

This book focuses on the three major types of maintenance programs in manufacturing facilities known as preventative, predictive, and preventative maintenance. It examines the advantages and disadvantages of these programs in layman's terms. Maintenance terminology can be quite complex, but the text in this book is written so it is easily understandable at any reader level.

Now that you understand the scope of this book, we can into discussion on the three major types of maintenance programs. However, before doing this, we need to discuss the planning of these programs in order to get a better understanding of how they are implemented.

Planning

Maintenance is very important in manufacturing facilities because it affects the livelihood of everyone in the organization. Machines need to run properly because broken machinery stops production...and production is the life-blood of manufacturing operations. In fact, production is major reason that most manufacturing plants exist. Based on this fact, it is relatively easy to understand the importance of maintaining machines in proper working order.

Maintenance programs need to be planned before they are implemented...regardless of whether the type of maintenance is preventative, predictive, or corrective. Planning starts by defining a purpose. Will the maintenance program be proactive, reactive, or projective? If the program is proactive, then the purpose is preventative maintenance. Machines will be serviced on a regular basis to prevent problems from occurring during production. If the program is projective, then the purpose is predictive maintenance. Machine failures will be predicted so they can be managed. If the program is reactive, then the purpose is corrective maintenance. Machines will be serviced when they break down.

Surprisingly, most manufacturers do not have predictive or preventative maintenance programs in place, choosing instead to take corrective action as needed. Money and time are two major factors that cause them to opt out of predictive or preventative activities because these resources are needed in other areas. This plan might work in the short term, but it can cause a wealth of machine problems down the road. However, regardless of the type of maintenance program chosen, the purpose of it needs to be defined.

Next the scope of the program needs to be outlined. Questions that need to be answered include:

Who will be involved?

What people are going to perform the work and what are their designate responsibilities? In a manufacturing facility, the scope might only be applicable to one department or it might encompass the entire plant. Some maintenance personnel are highly skilled while others are not, so specific roles need to be defined. There also needs to be a manager in charge who reports to upper management.

What machinery will be involved?

Manufacturers need to designate the equipment or machinery that is going to be maintained or repaired. Some machines are purposely left off of this list because they are only serviced by representatives of the companies that manufacture them. Other machines might be left off this list because they are located outside of the physical boundaries of the plant. For example, company owned vehicles (trucks, bulldozers, tractors, etc.) might not be serviced by maintenance personnel simply because it is more convenient and cost effective to use external sources.

What are the priorities?

Does one machine or department take priority over others in terms of repair? This is an important question because it gives maintenance personnel guidelines for allocating their time. Without a list of priorities, time can easily spent fixing machines that are not needed until a later time. The focus needs to be on equipment that is immediately needed for production.

What will be documented?

If utilized, will routine maintenance checks of machinery and equipment be recorded? Routine maintenance is done in many plants on periodic basis, and it is nearly impossible to remember the all of the dates and times that service was performed.

A document should be available that lists all routine maintenance checks and the frequency that those checks are performed. Frequency should be based on usage (volume), safety, and manufacturer recommendations. Documentation of routine maintenance serves three basic purposes:

It indicates when maintenance needs to be conducted

Documentation shows when a machine was last serviced, and when the next service is due. This eliminates the need to rely on memory. Technology today even allows for reminders of upcoming services that can be delivered directly to smart phones or other mobile devices.

It maintains warranties

Some warranties do not remain in effect if certain routine maintenance is not performed. For example, oil might need to be changed in a machine every 200 working hours or the warranty is not valid.

It provides information for authorities

OSHA, auditors, and government agencies all request routine maintenance information when investigating injuries, validating processes, or probing violations. IF this information is not provided, fines could be levied and customers could be lost.

What is the operating budget?

Like every other aspect of business, money plays a role in maintenance planning. Many manufacturers have budgets in place that limit expenses in maintenance departments. They allocate funding for various areas such as building and grounds upkeep, machinery repair, vehicle servicing, and building renovations.

Keep in mind that budgets are great for planning, but they do not always work for maintenance because there are often unforeseen circumstances. Machinery that is critical to production simply cannot wait until the next budget renewal...it has to be fixed now or the plant will not be able operate and orders will not be filled.

Now you understand the importance of planning for maintenance programs. Armed with this knowledge, it is time to move into the specific types of programs...starting with preventative maintenance.

Preventative maintenance

In general, preventative maintenance programs are implemented so future problems can be avoided. For example, tires on a car should be rotated every 8000 miles. This prevents the tires from wearing unevenly and needing to be replaced before the 40,000 mile life expectancy. In a manufacturing facility, preventative maintenance is also designed to prevent future problems from occurring. For example, working machine parts need to be greased on a weekly basis to avoid excessive friction that leads to damage.

Like most other aspects of business, preventative maintenance has benefits and drawbacks in manufacturing plants. These positives and negatives must be taken into consideration when deciding whether or not to implement a preventative maintenance program. The time, effort, and money invested into this type of program needs to have a payback in order to be justified…and that justification can only be determined by plant management personnel.

The following are some specific advantages and disadvantages:

Advantages

Below are some advantages of preventative maintenance programs.

Risk reduction

Preventative maintenance reduces the risk that there will be failures during production. It provides insurance that production quotas will not be interfered with by broken machinery or faulty equipment. This eliminates a major headache for management personnel and allows them to focus on other areas of their jobs.

Life expectancy

Every manufacturer wants machinery and equipment to last as long as possible in their facilities. This eliminates replacement costs that can be quite significant…especially when machines are designed for a single purpose. That being said, machinery and equipment are expected to last longer when a preventative strategy is utilized because periodic servicing helps maintain them in proper working order. In terms of life expectance, an ounce of prevention is truly worth a pound of cure.

Cost savings

When done properly, preventative maintenance easily justifies its existence economically. It helps (1) prevent maintenance personnel from fixing machines at a later date, (2) avert the need for external sources of repair, (3) avoid unnecessary down time and the cost of unproductive labor, and (4) eliminate sluggish equipment that slows productivity. Based on these four areas of cost savings, it is rather obvious that the payback for preventative maintenance programs can be substantial.

Energy

>This advantage goes largely unnoticed, but it needs to be noted. Machinery and equipment that are not serviced using preventive maintenance are often less energy efficient. They require more electricity or gas to function at desired levels, thereby increasing energy bills and wasting resources. Cost goes down and efficiency goes up when preventative maintenance programs are in place

Disadvantages

Below are some disadvantages of preventative maintenance programs.

Immediate costs

>There is an up-front cost for preventative maintenance programs. Personnel need to be hired and supplies need to be inventoried. Depending on the number of people hired and the scope of the program, this can be quite expensive…and all of the costs are accrued before the first product leaves the production line. Some companies cannot afford to put out the money, and others simply refuse to do it because they believe the cost is not justified.

Management

>This refers to maintenance people and preventative maintenance programs because both of them need to be managed. People require direction, and that directions needs to come from a supervisor. That supervisor also needs to make sure the program is followed and the work required gets done in a timely manner. Management of a preventative maintenance program is not a small task, and that makes it a disadvantage for manufacturers.

Volume changes

>When is preventative maintenance too much or too little? This question is difficult to answer, and it can create a problem for manufacturers. For example, a program requires maintenance personnel to replace the wheels on all smokehouse racks in a turkey processing plant every six weeks. This makes sense because racks could break down during production, causing downtime. However, this program does not account for seasonal volume shifts such as Thanksgiving (when production is at a peak), and the summer months (when production is very low). In reality, wheel replacement should be much higher around Thanksgiving and much lower in the summer months…but this is not the case because the preventative maintenance program calls for replacement every six weeks.

Value

Unfortunately, some business leaders believe preventative maintenance is a luxury. As a luxury, it is one of the first areas to undergo cuts when manufacturers are experiencing financial difficulties. From a cost saving perspective, this makes little sense because the money spent preventing problems is typically less than that spent repairing machinery or equipment that has failed. However, these leaders' thinking will likely never change because preventative maintenance is regarded as a "precautionary" expenditure that is difficult to tie to actual production downtime. If the value of something cannot be directly measured, then number crunchers in the manufacturing organization will push for its elimination during tough times.

Now you understand the advantages and disadvantages of preventative maintenance. This program is beneficial for many manufacturers because it keeps equipment and machinery operating at optimal levels. However, there are some up-front costs involved, and management needs to determine the real value of this program.

Next, let's move into a discussion on a type of program that uses logic and reasoning to assess maintenance needs. That program is known as predictive maintenance.

Predictive maintenance

This is the rarest type of maintenance program used by manufacturers. Essentially, machines and equipment are examined in order to predict when maintenance should be performed. Similar to preventative maintenance, predictive maintenance is implemented to avoid future problems. However, if done properly, this program costs less than preventative maintenance because service is only performed when justified. In other words, predictions are made about the potential failure of machines and service is performed just before those failures become reality. The goal is to avoid unnecessary maintenance expenses.

Predictive maintenance is also the most difficult type of maintenance program used by manufacturers. Timing is critical because service has to be performed before the failure with sufficient warning time must be provided. Techniques include observing machine performance, ultrasound, acoustics, thermal imaging, vibration analysis, and oil analysis.

The following are some specific advantages and disadvantages.

Advantages

Below are some advantages of predictive maintenance programs:

Maintenance time

This program predicts service needs. It falls under the preventative maintenance category, but service is only performed when it is justified by the potential for equipment or machine failure. This means less maintenance effort is needed for predictive maintenance, and the end result is a savings in time and labor.

Inventory

This is likely the least known advantage of predictive maintenance. Predictive maintenance does not require inventory of excessive machine parts because only the parts necessary for the program are kept in stock. Emergency spare parts stock no longer exists, and this results in cost and space savings.

Safety

Skill levels of personnel performing predictive maintenance are high because these individuals have undergone training and understand the machines and equipment they are servicing. As a result of their knowledge, safety levels increase throughout the plant. This safety is critical because many manufacturers work with hazardous chemicals or operate equipment that requires high pressure or temperature. It creates a win-win situation for employees and management because employees do not go through the pain and suffering associated with injuries, and management does not pay the costs associated with workers compensation.

Disadvantages

Below are some disadvantages of predictive maintenance programs.

Monitoring/testing costs

Specialized monitoring and testing devices are typically expensive. They are made for a specific task, so a higher price can be charged for them. The cost might be understandable, but it is also prohibitive for some manufacturers. They either cannot or will not spend the money necessary for the equipment, so the predictive maintenance program does not function properly.

Required skills

Every employee is not capable of performing predictive maintenance tasks. In fact, the vast majority of employees are not capable of performing these tasks because they require specific skills. In addition to having mechanical skills, people who do predictive maintenance often need training in electronics, hydraulics, or thermodynamics.

Environmental effects

Some manufacturing plants have conditions that are less than ideal for the monitoring or testing devices necessary for predictive maintenance. For example, food processors with wet or cold working environments might have problems keeping these devices working properly. The same goes for the high temperatures found in foundries or smelting plants. Along the same lines, paint manufacturers are likely to have corrosive chemicals that could do damage.

Regardless of the way the damage is done, monitoring or testing devices that are not working properly will not provide accurate information. This means calculations could be inaccurate, and the entire predictive maintenance program is jeopardized. Since it is difficult for some manufacturers to avoid destructive work environments, it is understandable why they choose not to implement this type of program.

Now you understand some of the advantages and disadvantages of predictive maintenance. This program is beneficial for many manufacturers because it analyzes machinery and predicts when it will fail. This information is then used to perform service before the failure occurs, while housing fewer parts and maintaining lower labor costs for maintenance personnel. However, people need specific skills to be employed as predictive maintenance technicians, and the type of work environment can affect the data collected.

Next, let's move into a discussion on a type of program that addresses machinery and equipment failures after they occur. That program is known as corrective maintenance.

Corrective maintenance

This is the most common type of maintenance program in manufacturing plants. It is a completely reactive program, and it is necessary because equipment and machines will break down at some point. Machines cannot be expected to run forever, and constant use at full capacity typically shortens that life span.

Unfortunately, corrective maintenance is often the only type of maintenance program available in a manufacturing facility....with no predictive or preventative programs to support it. Many times this is due to cost because smaller manufacturers cannot afford to sacrifice the resources necessary to set up predictive or preventive programs. However, sometimes corrective maintenance stands alone simply because organizations do not want to invest the necessary time and effort to establish other programs. Corrective maintenance does wonderful things for equipment and machinery repair, but it needs help. Without some type of support, corrective maintenance programs can become very expensive in a relatively short period of time. This adds stress to the jobs of maintenance personnel and managers, and they might start looking for jobs elsewhere.

Advantages

Below are some advantages of corrective maintenance programs.

Initial investment

Corrective maintenance does not require the planning, time, or effort required for preventative and predictive programs. This is because corrective maintenance does not address problems before they occur; it simply reacts to issues at the time of failure. This is advantageous for manufacturers because they save on resources. In short, there is a savings on initial investment for manufacturers that choose to corrective maintenance as their only maintenance program.

Expenses

Corrective maintenance delays expenses. These expenses include the services and checks performed under preventative and predictive maintenance programs. This means machines and equipment can function for extended periods of time with little or no maintenance. This strategy is particularly beneficial for manufacturers looking for short term return on investment, such as that expected from machinery or equipment needed for a specific purpose. For example, a toy manufacturing company might need a machine to stitch stuffed animals for two months until they implement an entirely new process. Management does not want to put time and money into maintenance of this machine unless it completely fails. Even if the machine breaks down, it will be "quick-fixed" or temporarily repaired because it will not be needed for the long term. In this case, the short term return justifies corrective maintenance being the only program in effect.

Profitability

Profit is a major reason that most manufacturers are in business, and higher profits can be made by organizations that prefer to react to maintenance issues as they occur. They are willing to forego using any type of predictive or preventative programs in order to make more money. Savings from labor, supplies, and parts all lead to higher profitability...and a happier management team.

Disadvantages

Below are some disadvantages of corrective maintenance programs.

Predictability

As most manufacturers are aware, this is likely the biggest disadvantage of a corrective maintenance program. Maintenance personnel do not know when equipment or machinery will fail, and that can cause a variety of different problems. For example, parts might need to be ordered, thereby delaying the repairs necessary to get production running. Additionally, outside service might need to be called in for issues that cannot be resolved by plant personnel...and that service is typically quite expensive. When these problems mount, the cost of the corrective maintenance program can far exceed that of a program that had preventative measures in place.

Efficiency

Efficiency is important for every production oriented facility because increasing it helps keep costs down and maximizes productivity. Unfortunately, corrective maintenance programs do little for efficiency. The major goal of corrective maintenance is to keep equipment and machinery operating, but optimal levels of operation are not necessarily part of that goal. When optimal levels are not achieved, equipment and machinery do not reach their potential...and the resulting lack of efficiency causes a decline in productivity.

Urgency

As has already been stated in this book, corrective maintenance programs do not prevent problems. This makes the likelihood of problems much more probable, and those problems need to be addressed immediately when they affect production. As might be expected, most machinery problems hinder production, so repairs need to be made with no time to waste. Unfortunately, this type of environment creates a wealth of stress for maintenance personnel and managers...and that is why urgency is a negative associated with corrective maintenance.

Now you understand some of the advantages and disadvantages of corrective maintenance. This program is beneficial for many manufacturers because it minimizes expenses and raises profitability. However, repairs necessary for corrective maintenance difficult to predict, and there always tends to be a sense of urgency.

The next section summarizes preventative, predictive, and corrective maintenance programs for a better understanding.

Summary of the three types

The three major types of maintenance programs have now been described. However, their specific applications might still be a little difficult to understand unless they are compared side-by-side. Based on this thinking, a brief and concise summary is as follows:

Preventative maintenance programs are proactive, predictive maintenance programs are selectively proactive, and corrective maintenance programs are reactive.

Examples of the work performed by each type of program are as follows:

Preventative maintenance – Greasing or lubricating working machine parts, changing oil in machinery
Predictive maintenance – Measuring the amount of vibration on machines, searching for gas leaks in machinery
Corrective maintenance – repairing machinery after it has broken down, replacing broken safety covers machines

In a nutshell:

Preventative maintenance programs administer a wide variety of services that prevent failure, predictive maintenance programs do specific testing to determine services that prevent failure, and corrective maintenance programs perform services after failure has occurred.

Based on what is written in this book, it is rather obvious that maintenance programs are necessary to keep equipment and machinery functioning properly in manufacturing facilities. That being said, existing maintenance programs need to be continually updated and improved upon. This might not be easy, but it is important...and it could affect the survival of some manufacturers.

Summary

Maintenance personnel are essential for any type of production oriented operation. They assemble, diagnose, repair, and monitor the equipment and machinery needed to fill orders and satisfy customers. In terms of manufacturing, maintenance departments are the glue that holds facilities together.

This book focuses on maintenance programs in manufacturing. First it examines the planning that takes place before maintenance programs are implemented while discussing the people, priorities, and documentation involved. Then it analyzes preventative, predictive, and corrective programs through description and an exploration of their advantages and disadvantages. The text is educational and informational, and it is written for easy reader understanding at all levels.

Congratulations! You now understand more about preventative, predictive, and corrective maintenance...three important types of maintenance programs used by manufacturers.

www.ingramcontent.com/pod-product-compliance
Lightning Source LLC
Chambersburg PA
CBHW070412190526
45169CB00003B/1218